Contents:

Chapter 1: How We Are Amazingly Made ………..

Chapter 2: Why Is It "Super-Intelligence"? ……… 13

Chapter 3: Why Not "Evolution"? ……………... 17

Chapter 4: God In Our Governments ………….. 21

Chapter 5: Why Is Evolution Still Being Taught? ... 25

Chapter 6: Why Does Our Creator Care For Us? ... 27

Chapter 7: What Should Educators Do With This? . 29

Index: ……………………………………….. 31

Contact Us: ………………………………….32

Introduction:

Have you ever thought of all the amazing and careful works our Creator performs for you every day? It is worth a pause, just to smell a rose, or watch some birds, or hug a friend, or eat some fruit and enjoy good food.

For most of us, these things are so easy and reliably available we usually take them for granted, but none of this just happens. Each "Life" takes a huge amount of careful work by our Creator, the God of our nation.

The study of this brilliant work is a new subject called "atomic biology", and it is fascinating.

Hopefully, most teachers and professors want to provide accurate information to their students. Ideally, more education leaders will insist that best explanations are taught, especially in the sciences. This means allowing students to "follow the evidence wherever it leads," but enforced teaching of evolution-only as the cause of life, disallows teaching that super-intelligence is essential to construct our cell-parts and us. Conscientious education leaders are needed to fix this problem.

"If it could be demonstrated that any complex organ could not possibly have been formed by numerous successive slight modifications, my theory would absolutely break down." [1] -- Charles Darwin

This summary will show:
1. the absolute essentiality for Super-Intelligence, far beyond mankind's level, in building all living entities; and
2. the enormous amount and speed of the intelligent works performed for us in our bodies every second of every day; and
3. the phenomenally careful precision of the brilliant construction work using atoms in building each of our cell-parts, cells, and the total "us".

These are all works that "evolution" is incapable of performing because, by definition, it has no intelligence to use.

For many reasons that you will see in this booklet, evolution should be ruled out as the true cause of life.

References:
[1] Darwin, Charles, *On the Origin of Species by Means of Natural Selection,* 1st edition, 1859, p. 189, John Murray, London, Eng., available online from Darwin-online.org.uk.

Endorsements for the full First Edition of
"Darwin's Replacement"

"The need for a super-intelligent force to create and sustain living things is well set out and without question. I also appreciated all the research that was done to demonstrate the historical importance and recognition of God in the nation lives of the four nations you selected."   -- George Matzko, Ph.D.

"*Darwin's Replacement* is a well-written review of the enormous complexity of all life written from the perspective of the molecular biochemical foundation. The complexity is more than amazing. Most people enjoy learning about amazing feats, and thus the popularity of 'Ripley's Believe It or Not' and similar books. Our body and its complexity are so familiar and work so well for most of us that we often take it for granted. Mr. Rogers' book helps us to realize that we are all walking miracles; and understanding how it functions is both awe-inspiring and helps us to appreciate the body we all live in while on this Earth."   -- Jerry Bergman, Ph.D.

"*Darwin's Replacement* provides credible assessment of the weaknesses within mainstream evolution theory, and proposes a reasonable, evidence-based alternative for the origin and development of life.   -- Nicholas Comninellis, M.D.

"ATOMIC BIOLOGY promises to restore the true foundations of science back to the realm of observation. Current forays into metaphysical speculation and presupposition by 'experts' seem to have caused division, confusion, and misunderstanding in the scientific conversation."

-- Jack Taylor, Ph.D.

"My recommendation would be to rework the book as a series of 1-to-2-page studies for adult Sunday school classes and/or Christian High School classes. It could be useful as an introduction to the topics, in that format."- David Snoke, Ph.D.

"Your book demonstrates the amazing complexity of life, starting with even the simplest cell, and the numerous conditions needed to sustain life. That all this could be the result of blind, random evolution is highly implausible, and statistically, virtually impossible. Hence, as your book concludes, this points to a super-intelligent Creator. Your book also notes that the USA, the UK, Canada, and Australia were all founded on submission to the Christian God, and urges those countries to return to acknowledging God, also in science classrooms. I heartily agree with all this." -- John Byl, Ph.D.

"There are only two possibilities for the existence of life: accidental or purposeful. Using science and mathematics, Atomic Biology proves beyond a shadow of doubt that life cannot be accidental. Then the book shows that the only being capable of the creation of life and its orchestrated maintenance is the historical Omniscient, Omnipotent, and Omnipresent Triune God of the Bible and our Nation." -- Sharon E. Cargo, D.V.M.

"Hi, (Reality R&D). I recently bought the book and am reading it slowly and out loud to myself so that it sticks, but can I say that when I heard Mr. Rogers on Vision I knew that this was a book I'd been waiting for a very long time. I honestly can't put into words how exciting this is for me to finally have something I can refer to when discussing creation and not sound like a loony." -- Linda Houston, Australia

DARWIN'S REPLACEMENT SUMMARY EDITION II

HOW WE ARE AMAZINGLY MADE
AND CARED FOR BY OUR SUPER-INTELLIGENT CREATOR, THE GOD OF OUR NATION.

Dedication

To Our Super-Intelligent and Caring Creator, Credit for Life Where Credit Is Due.

Acknowledgements

Our respect and gratitude go to the forty-five scholars whose work has helped in the discovery and development of this life-science called "atomic biology." Our Creator has probably been using this science since the beginning.

We are also grateful to the encouragers and endorsers who have helped to keep the work moving forward.

LRG

LIFETIME REFERENCE GUIDES INC.

P.O.Box 51613 RPO Park Royal
West Vancouver, BC, Canada V7T 2X9
www.lifetimereferenceguides.com

Copyright © 2020 by Lifetime Reference Guides, Inc.

Cover design by ArneeonMedia.com with Shutterstock images.

All rights reserved.

No part of this publication may be reproduced in any form by any means, electronic or mechanical, including photocopying, recording, information browsing, storage, or any retrieval system, without specific written permission from the publisher.

ISBNs
978-1-9992097-6-6   (eBook)
978-1-9992097-5-9   (Paperback)

Printed in Canada

Science
Education
Government
Biology

Chapter 1: How We Are Amazingly Made

Charles Darwin theorized in his 1859 book, *On the Origin of Species,* that a common ancestor existed for all living things. However, to this day there has been no plausible explanation for how that necessarily complex original ancestor came into existence without intelligent help. The construction of even the simplest cell is now known to be so utterly complex that even a growing number of evolutionists disagree with Darwin's theory.

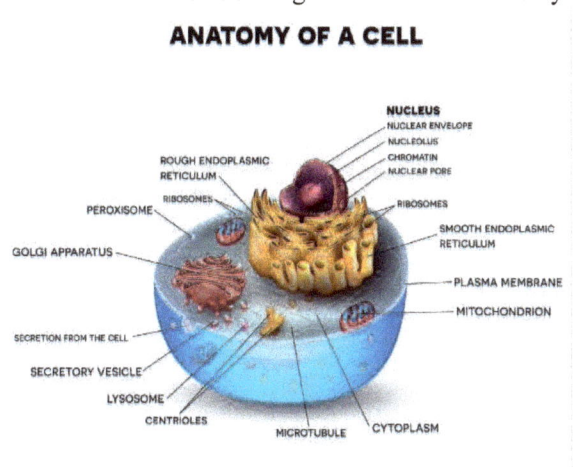

Shutterstock – Tefi

It is this brilliant design and construction of cell-parts, cells, and living entities like us, that has compelled scientists to move away from their former belief that "evolution" is the cause of all life.

Here are some of the many amazing factors that illustrate the brilliant design and construction works for life by a super-intelligent and caring force:

* A 150-pound male is built with about 100 trillion cells constructed including over 200 cell types;
* Each one of our approximately 20 trillion red blood cells is constructed from about 280 million hemoglobin molecules,[1] and each of those molecules consists of about 10,000 atoms.[2] So, each one of our red blood cells consists of about 2,800 billion correctly selected, counted, and assembled atoms;
* About 2.3 million new red blood cells are produced every second,[3] 24/7, to replace those that are worn-out in this average 150-pound male (i.e. about 6400 quadrillion atoms/second). You can estimate how many atoms per second are being selected and assembled to make your red blood cells, using your relative weight. The constant replacement of our blood cells and most every other cell type in our body is a huge and complex task performed for every human by our caring Creator.
* Our many cell organelles must be constructed to exacting specifications. For example, our four DNA bases are made precisely to the following formulae:

<u>A</u>denine - chemical formula $C_5 H_5 N_5$
<u>G</u>uanine - " " $C_5 H_5 N_5 O_1$
<u>C</u>ytosine - " " $C_4 H_5 N_3 O_1$
<u>T</u>hymine - " " $C_5 H_6 N_2 O_2$

Notice how carefully the atoms have to be selected, counted, and assembled. One atom off is likely lethal. For Adenine, $C_5H_5N_5$ is required. $C_5H_5N_6$ will not work, nor will $F_5H_5N_6$. The required atoms are picked up from our digestive system and carried in our blood stream to where they are needed for constructing new cell parts. For DNA, over three billion pairs of these bases must be precisely assembled and arranged in special sequence to construct the hardware and assemble the functioning DNA and RNA software instructions required to operate up to 40 molecular machines in each one of most of our cells;

* How about our amazing eyesight?
Even Mr. Darwin said, "*To suppose that the eye with all its inimitable contrivances for adjusting the focus to different distances, for admitting different amounts of light, and for the correction of spherical and chromatic aberration, could have been formed by natural selection, seems, I freely confess, absurd in the highest degree.*"[4]

Then he wrongly speculated on how it might just happen without intelligent guidance.

Human Eye Anatomy

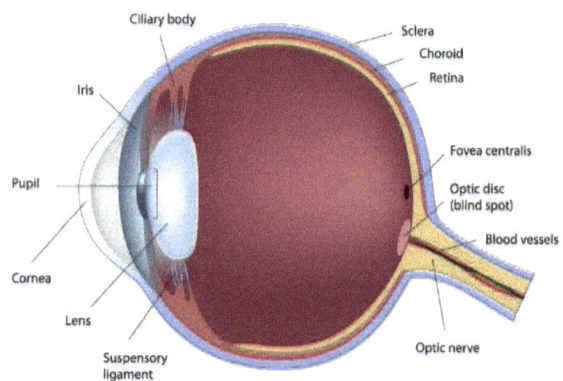

Alila Medical Media/Shutterstock

* Then there are all our other amazing senses including hearing, tasting, smelling, and touching (awesome and reliable but so easy to take for granted);

* How about our incredible brains? Prof. Paul Reber estimates our brain storage capacity is equivalent to about three million hours of recorded TV shows.[5] Our brain is constructed without any intelligent work and care? Let's be real.

<u>Magnified Image of a Cell Membrane Portion</u>

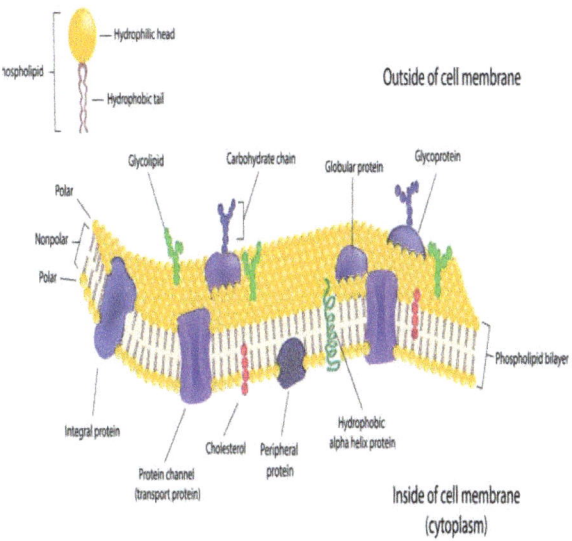

Kallayanee Naloka / Shutterstock

Notice how complex it is to build even this small portion of the membranes encasing each of our cells. What a phenomenal amount of careful works with atoms are essential. (No charge).

The Simple Solution to the God vs Evolution Question

Either evolution is the cause of life, or it is not.

Proof now exists to show that Super-Intelligence (far above the human level), is essential to construct our various cell-parts, cells, and us.

Evolution, by definition, has no intelligence to use, therefore, it is falsified as the cause of life.

Evolution should no longer be taught as the cause of life.

Unfortunately, you will still have to say that "evolution does it" on your exams for a while longer, but you can also remember the above factors regarding how much your Creator cares for you.

You should not fight your professor over this, but you can decide what is logically true.

References:

[1] Tortora, G. J., *Principles of Anatomy and Physiology,* John Wiley & Sons, New York, NY, 2008.

[2] Perutz, Max, *Science is Not a Quiet Life: Unraveling the Atomic Mechanism of Hemoglobin,* World Scientific, Hackensack, NJ, 1997.

[3] Pallister, C. J., *Haematology: Biomedical Science Explained,* Butterworth-Heinemann, Burlington, MA, 1999.

[4] Darwin, Charles, *On the Origin of Species by Means of Natural Selection,* 1st edition, 1859, p. 186, John Murray, London, Eng., available online from Darwin-online.org.uk.

[5] Reber, Paul, "What Is the Memory Capacity of the Human Brain," *Scientific American,* May/June 2010, www.scientificamerican.com/article/what-is-the-memorycapacity/.

Chapter 2: Why Is It "Super-Intelligence"?

In 2016 the three Nobel Prize Winners in Chemistry, J.P. Sauvage, J.F. Stoddart, and B.L. Feringa, showed, unintentionally, that mankind has nowhere near enough intelligence to build even the simplest molecular machine in our cells. They received this prestigious prize after 33 years of work to make a few simple molecular machines. The best they can make is, relatively speaking, almost infinitely more simplistic than the simplest ones built for our new cells every day.

This shows that "Super-Intelligence" (far more than mankind has) is essential for building our molecular machines for cell-parts, cells, and ultimately us.

The evolutionists who say that "natural mechanisms" do this work, invite a much larger problem for their theory: these "natural mechanisms" would first have to be constructed with far more intelligence and better equipment than the Nobel Winners had for use. Without intelligent help, how probable is that?

The only *Super-Intelligent Cause* known to mankind is the God of our Nations (USA, UK, Canada, and Australia). The God of "In God We Trust," "One Nation Under God," "God Save The Queen," "God Keep Our Land Glorious and Free," the God of "Thanksgiving Day", "Christmas", and "Easter" -- that God.

He constructs our cell-parts and our body using atoms from our foods which He made with the same atoms from the dust or soil in gardens, fields, and orchards.

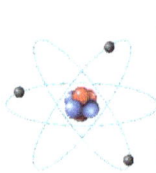 

Shutterstock/Azure

Shutterstock/Olga Lyubkina

Van Wedeen and L.L. Wald of the Martinos Center for Biomedical Imaging Human Connectome Project, state that "The brain's many regions are connected by some *100,000 miles* of fibers called white matter, enough to circle the Earth four times."[1] Making our brains out of "dust" requires Super-Intelligence.

Jeff Lichtman, a neuroscientist and professor at Harvard, is studying brain compositions. He was interviewed by Carl Zimmer who wrote in *National Geographic,* February 2014, "So far the largest volume of a mouse's brain that Lichtman and his colleagues have managed to re-create is about the size of a grain of salt. Its data alone totals a hundred terabytes, the amount of data in about 25,000 high-definition movies." They also related that a mouse's brain contains about 70,000,000 neurons and a human brain contains about 1,000 times that number.[2]

Douglas Axe, Ph.D., is an engineer-turned-molecular-biologist and director of the Biologic Institute. In his recent book, "*Undeniable: How Biology Confirms Our Intuition That Life Is Designed,"* he reminds us that, "*The human brain is different.... Being the most remarkable component of the human body, it is arguably the most outstanding physical invention ever to exist."*[3]

This phenomenal computer, our brain, had to be super-intelligently built of atoms from the soil, air, and water with God's phenomenal two-step process: atoms in the soil are built into food, then those same atoms in our food are used to construct our brain cells.

Where else could these atoms come from, and who else has the super-intelligence to assemble them in such a phenomenal manner?

Then, of course, there is the matter of adding "Life" to these inanimate atoms. (More Super-Intelligence required).

God's "Evolution" ability is shown here in transforming a crawling worm into a beautiful flying butterfly in 14 days, not millions of years.

Stephen Russell Smith Photos/Shutterstock

Evolutionists might say this is just "metamorphosis" but the name cannot do the work.

References:

[1] Wedeen, Van, and Wald, L. L., in article "Secrets of the Brain" by Carl Zimmer, *National Geographic,* February 2014, p. 34.

[2] Lichtman, Jeff, in article "Secrets of the Brain" by Carl Zimmer in *National Geographic,* February 2014, pp. 39, 43.

[3] Axe, Douglas, *Undeniable,* Harper One, New York, NY, 2016, p. 259.

Chapter 3: Why Not "Evolution"?

As "evolution" has many definitions, we must define the word so the reader and author can be 'on the same page.'

The definition of "evolution" that we, at The Atomic Biology Institute, are using includes the theories of evolution currently being taught in our public schools, colleges, and universities, including Darwinisms, Neo-Darwinisms, and macro-evolution. They include the theoretical concept of a "common ancestor" formed by a chance assembly of atoms billions of years ago. This imagined "common ancestor" had to come into existence by a chance assembly of atoms that was complex enough to have life, to function, find nourishment and digest that for survival, and reproduce with improvements.

The mathematical odds against this happening are so enormous as to make it realistically impossible.

No one is able to adequately explain how such a complex original ancestor could have come into existence without intelligent work. In reality, there is our "common designer and builder".

For the imagined "common ancestor" to have the intelligence to assemble an offspring with even more sophisticated features is such a ridiculously improbable assembly job that it should put the whole theory of evolution into the scientific trash bin.

It is the highly complex construction work that goes into even the simplest living entities that is turning an increasing number of scientists away from "evolution" as the cause of life.

Cambridge University Ph.D., Stephen C. Meyer, stated in his book, *Signature In The Cell,* that "The simplest extant (still surviving) cell, *Mycoplasma genitalium* – a tiny bacterium that inhabits the human urinary tract – requires 'only' 482 proteins to perform its necessary functions and 562,000 bases of DNA (just under 1,200 base pairs per gene) … .

" Based on minimal-complexity experiments, some scientists speculate (but have not demonstrated) that a simple one-cell organism might have been able to survive with as few as 250 to 400 genes."[1]

Can anyone honestly believe that a living entity even this small could be constructed by chance using no intelligent help or work? Let's be real.

We have already pointed to the Nobel Prize Winners' evidence showing that super-intelligence is essential to build the molecular machines required for our cells to function. Evolution claims that intelligence is not required to produce life. All that is required is time, chance, and the traditionally accepted four forces that hold the universe together, namely, gravity, electro-magnetism, and the strong and weak nuclear forces.

In addition to no evidence of a common ancestor, and evolutionists' claim that no intelligence is necessary for building cell-parts and cells, are the following factors:

* The atoms our cells are made of do not have the ability to move themselves into their precise position in a cell's organelles. This requires a super-intelligent cause to find, select, count, and precisely assemble them in sequence;
* The fossil record shows no evidence of clear transitional forms for the major categories of life. The few fossils claimed to be transitions are dubious at best. This was a major concern for Darwin as well;
* *"Dead dogs don't bark,"* which is to say that although all the required atoms, molecules, and cells are precisely built and placed into their correct position for eyes, ears, teeth, brain, legs, heart, and so on, *without the super-intelligent breath-of-life, those atoms, molecules and cells, or the dog, are not going to move one millimeter.* This God-given breath-of-life is crucial to every living entity. When removed, the entity's life ends.
* All living entities are built of cells containing DNA. DNA is like a computer software program which uses a very sophisticated, highly complex coding system to assist in the multiple and varied functions that cells have to perform in order to help keep a living entity alive. Complex, intelligent, functional codes,

like those in DNA, cannot be programmed without an intelligent programmer.
* Therefore, Evolution, having no intelligence to use, is falsified as the cause of life.
* There are many factors that must be intelligently 'tuned', highly regulated, and crucially consistent in order for our beautiful planet to function, and for living entities to exist. Random, uncontrolled, conditions would quickly lead to extinction of all life.

Here are just a few conditions which life must have in order to exist:

1. Temperatures must not be too hot or too cold;
2. Sufficient water must be available;
3. Sufficient food must be constantly constructed;
4. An intelligent force (God, the only possible one) is needed to build cells from atoms;
5. God is needed to "breathe life" into these cells;
6. Adequate sunlight must exist for life to exist;
7. The atmosphere must be just right;
8. Gravity must not be too weak or too strong;
9. Electricity flow must be controlled in our body;
10. A super-intelligent force is needed to keep all of these necessary factors (and more) in balance.

Reference:

[1] Meyer, Stephen C., *Signature in the Cell,* Harper Collins, New York, NY, 2009, p. 201.

Chapter 4: God In Our Governments.

In this writing, we mean the governments of the USA, the UK, Canada, and Australia.

Many decades ago, our forefathers and founding fathers intuitively understood that living entities including livestock, food, siblings, children, and themselves, were all made by an intelligent cause whom they called "God," their "Creator," and their "Provider."

They also found truth and internationally beneficial wisdom in His Word, the Holy Bible.

Out of respect and appreciation, they implemented special days of recognition for God's faithful works on their behalf including Thanksgiving Day, Christmas, and Easter.

They highly recognized God in their Declarations, Justice Systems, Constitutions, Pledges, Anthems, and more.

They put His name on Public Buildings, War Memorials, Currency, and more.

They prayed to Him in times of personal, national, and international distress then, as leaders do today.

He is very patient and forgiving with us, but only to the point where He is purposefully disrespected or angered because praise is given somewhere else for the enormous work He performs for each of us every second of every day.

History shows what happens to people and nations who turn their focus against Him.

Our students have the right to be taught "Why God is Highly Recognized by Their Government."

Now the life-science of atomic biology gives scientific reasons that back up the Governments' position of recognizing God.

Fortunately, in the USA, recognitions of our Creator and appreciation of His works are regaining prominence as states pass legislation to reinstall the national motto, "In God We Trust," back into public schools. Also, as skepticism grows against Darwinisms, some states are seeking an alternative science to teach regarding the cause of life.

We propose that "Atomic Biology" is a logical candidate to be considered as this significant Darwinism alternate.

As one who lived during the "Happy Days" of the 1950's, when God and His advice were widely taught, honesty and trust were honorable and practiced traits.

Those of us who know the benefits of our Creator's influence, are aware of what we have lost by the widening disrespect and ignoring of His blessings.

If we want God to bless America and other nations, we need to heed His advice and appreciate His enormous works on our behalf.

In addition to replacing the national US motto, "In God We Trust," into the halls of education, it would be good to explain "Why" to the students.

Reinstalling "The Ten Commandments" at the same time would provide some lifetime wisdom for our students to live by. The 10 were given to us by the God of our Governments for the benefit of our people.

A major goal of The Atomic Biology Institute is to explain some the multitude of reasons why we can trust and appreciate God and His wonderful works for each one of us.

It would also be helpful for students to understand their enemy who can tempt them, pushing some who give-in to temptations, into <u>deep</u> trouble. A thorough understanding of how temptation can be terribly destructive, would be an important lesson for all individuals to be taught.

Often temptation can make some action look exciting or fun or instantly gratifying, but the consequences can be devastating in the short term or even for a lifetime. There are probably no more devastating temptations than the abuse of alcohol, drugs, or sex.

A brief but graphic course on "choices and consequences" near the beginning of every school year, could save many individuals a lifetime of grief.

Governments, through their education departments, have a hugely significant opportunity to positively influence the lives, gratitude, enjoyment, personal health, ambulance costs re: over-dosers, health-care costs, policing costs, and heart-aches, for multi-millions of citizens and taxpayers.

A huge part of education is the imparting of practical wisdom and a moral compass, in addition to practical knowledge for our students. Unfortunately, evolution has greatly eroded the morals of the nations where it has been allowed to become dominant.[1]

The topics of "Choices and Consequences" will be discussed thoroughly in our textbooks.

For clarity regarding the various interpretations of "separation of church and state," we must point out that God is not 'the church'. The church can be a building or a group of people who are Satanists, New Agers, Catholics, Protestants, Mormons, Jehovah's Witnesses, cultists, or others. No church can create any living thing. God is a highly regarded part of some churches, just as He is a highly regarded part of our governments, but He is definitely not 'the church'.

References:

[1] See Dr. Jerry Bergman, *How Darwinism Corrodes Morality: Darwinism, Immorality, Abortion and the Sexual Revolution*, Joshua Press, Kitchener, ON, Canada; 2017; Dr. Jerry Bergman, *The Darwin Effect. Its Influence on Nazism, Eugenics, Racism, Communism, Capitalism & Sexism*. Master Books, Green Forest, AR, 2014

Chapter 5: Why Is Evolution Still Being Taught?

The first reason is the enforced exclusive teaching of the theory of evolution as the cause of life. No criticism is allowed. This is truly **anti-science** because it forcibly eliminates the basic principle of scientific discovery, i.e. encouraging exploration and following the evidence wherever it leads.

Restrictions of this type stifle the development of the best explanations, best solutions, and best products.

These 'evolution-only' rulings should be illegal, but in reality, the opposite is currently true. Teachers and professors who have dared to suggest that there may be some intelligence in the designing of living entities, have been severely repremanded.[1].... more anti-science.

However, in growing numbers, brave scientists are daring to publicly declare their skepticism of the dogma of evolution. It is bizarre that in democracies where freedom of thought and speech have been fought for, and died for, that there should be these penalized restrictions in science, a field of great potential for tremendous good.

The second reason is that the religion of Darwinism has been allowed to dictate restrictions onto other religious belief systems. This is a conspiracy of major negative significance to our society in general. They have been allowed to use our public educational institutions as their "churches" and teachers to preach exclusively their doctrine of 'evolution-only is the cause of life – no God allowed.'

Three generations of enforced teaching of Darwinism as the cause of life, has muddied the image of the God of our nations, our Creator and food Provider on whose principles our great nations are founded.

History shows that ignoring God and His advice leads to trouble for the nations, including moral decay and the strife that comes with it: disrespect for law, depression, anxiety, addictions, family breakdown, greed, and crime plus all the related costs. All these problems come from rejecting His wise advice, including particularly the ten commandments in His 'operator's manual,' the Bible.

It stands to reason why there is less strife for people who use this great advice.

Now that Super-Intelligence, (far greater than that of mankind), has been 'proven' essential for the design and construction of cell-parts, cells, and living entities, including us, it is time to bring the God of our nations back to our students.

References:
[1] See Jerry Bergman, *Censoring the Darwin Skeptics*, Leafcutter Press, Southworth, WA, 2016;
Jerry Bergman, *Slaughter of the Dissidents,* Leafcutter press, Southworth, WA, 2012

Chapter 6: Why Does Our Creator Care For Us?

Because our Creator's heart, mind, and capabilities are so much greater than ours, it is virtually impossible to thoroughly understand Him. However, He says He made us in His image and amazingly, we can communicate with Him at any time.

The wisdom He provides for us to live by, is for our most enjoyable life.

We can observe His magnificent works all around us in His construction and maintenance of the beautiful people, foods, pets, flowers, birds, tropical fish, trees, mountains, waters, sun, moon, and stars.

Less obvious but equally as profound are His essential controls of gravity, electromagnetism, the strong and weak nuclear forces, electricity, light, and temperatures.

This God-based life-science of atomic biology reveals many of the details of His care for us every second of every day.

For all these awesome works done for us, there is no charge. Like the loving parent that He is, He keeps giving and giving to us, whether we thank Him or not.

But, like any parent, He does expect some appreciation and recognition before we leave this world. He gives us a lifetime to acknowledge His enormous work and care for us. It is best to have him as a close and appreciated generous friend, at all times.

It is logical and understandable that He can become angry when someone or something else is given the credit for all the constant and faithful work He performs for each of us. The best-selling *Holy Bible* and other history books describe the troubles that come upon peoples and nations that disrespect Him and give credit for His enormous works to other causes like idols or evolution.

Among the many benefits and values of knowing our awesome Creator, Provider, and Maintainer is in cherishing the 'proof' of His obvious works, love, and care for each one of us.

This knowledge and understanding will be particularly beneficial for those students who learn it. The wisdom He provides in His best-selling 'operator's manual' has been of great benefit and value for centuries to those individuals and governments that have used it.

No scientist should be forced to pretend that there is no intelligence involved in the design and construction of living entities, including us.

One wonders how so many scientists who should know better, have been so obviously deceived or intimidated. There are, of course, many who just do not want to acknowledge or appreciate their essential, super-intelligent Creator and Provider but they should not be allowed to keep the evidence away from others, especially students.

Chapter 7: What Should Educators Do With This New Science of Atomic Biology?

As with all new concepts in science, atomic biology should be investigated impartially by individuals or committees that each group of educators deems reliable.

In this case, the term "impartial" is key because the topic of how living entities are constructed raises some emotional reactions, and emotions do not determine accuracy in science.

True science follows the evidence wherever it leads.

Through our three decades of research into what physical works with atoms have to be performed at each cell-construction site, we uncovered some very significant evidence. Our investigation of the 33 years of works by the three 2016 Nobel Prize Winners in Chemistry, provided the clinching evidence of the essentiality for a Super-Intelligent Cause in designing and building the 40 different kinds of molecular machines used in our 200+ different kinds of cells.

This was not the intent of the Nobel Winners, but it was a significant finding that derived from their work. Although they managed to build a few tiny molecular machines, the best units they could construct after 33 years of research and development are almost infinitely more simplistic than the simplest machines built for our new body cells every day of the week.

What they unintentionally 'proved' is that it takes far more intelligence than mankind has in order to build the amazing molecular machines required for our cells.

As, by definition, evolution has no intelligence to use, evolution is now shown to be falsified as the true cause of life.

As mentioned, when we asked evolutionists how they think the complex molecular machines for cells are made, those we have talked to declare that "natural mechanisms" do it. However, these natural mechanisms would have to be built containing far more intelligence and better equipment than the Nobel Winners possessed. Really?

We have identified seven other principles and eighteen other essential works necessary to cause life that evolution is incapable of performing. These are listed in our study/textbook also titled, *How We Are Amazingly Made*.

Our suggestion for education leaders is to first investigate atomic biology for accuracy, logic, and probability as the best explanation for the cause of life. Then obtain permission to allow teachers and professors to show students the evidence of how their awesome cell-parts and cells are carefully constructed by a Super-Intelligent and caring Creator, the God of their nation, in whom they can trust.

These booklets and study/textbooks are available by contacting us at www.atomicbiology.com .

INDEX:

A *Amazing 1,2,5,7-10, 30*
Anti-science 5,25
Atomic biology 3,17, 22,27,29,30
Atoms 6,8,9,13,15, 17-20,29
Axe, Douglas 15,16

B *Benefits 22*
Bergman, Jerry 1,24, 26
Brain 10,11,14-16,19
Brilliant 6,7
Butterfly 14
Byl, John 1

C *Care 5,6,9,10,23, 27, 30*
Cargo, Sharon 1
Cause-of-life 1,6,8, 17,19,21,22,24-26,30
Cell-parts 6,7,13,18, 26,30
Common ancestor 7, 17,18
Comninellis, Nick 1

Complex 1,2,5,7,17-20,30
Construction 6-8, 17, 26,27,29
(Contact us 32)
Counting 8,9
Creator 2,3,5,21,22, 26,27,30

D *Darwin, Charles 1, 5-9,11,17,19,22,24, 25,26*
Design 7,8,15,25, 26,29
DNA 8,9,18,19

E *Essentiality 6,13,18, 26,27,29,30*
Evolution 1,2,5-8,13, 16-19,24,25,28,30
Eye 9,19

F *Feringa, B.L. 13*
Fibers 14
Fossils 19

G *God 1,2,13,16,19, 21,26,27,30*

I *Incapable 6,30*
Institute 15,17,22

Intelligent 1-3,6-10, 13,15,17-21,25,29,30
L *Lichtman, Jeff 15,16*
M *Matzko, George 1*
Meyer, Steve 17,20
Machines 9,13,18, 29,30
P *Pallister, C.J. 11*
Perutz, Max 11
Q *Quadrillion 8*
R *Real 10,18,25*
Reber, Paul 10,11
Red blood cells 8
S *Snoke, David 2*
Sauvage, J-P. 13

Stoddart, J.F. 13
Super-Intelligence 1-3,6-10,13,15,17-21,25,29,30
Speed 6,
T *Taylor, Jack.*
Tortora, G.J. 10
True 1,6,25,29,30
U *Undeniable 15,16*
V *Values 27*
W *Wald, L.L. 14,16*
Wedeen, Van 14,16

Contact Us:
For information on our textbooks, study books, courses, speakers, discounts, donations, participation, or other questions, please see our website: www.atomicbiology.com

Your participation is invited to help bring our Creator, the God of our nation, back to our students.

www.ingramcontent.com/pod-product-compliance
Lightning Source LLC
Chambersburg PA
CBHW061732070526
44583CB00024B/3111